What Helps People See?

Color to find the hidden picture.

• = yellow •• = green

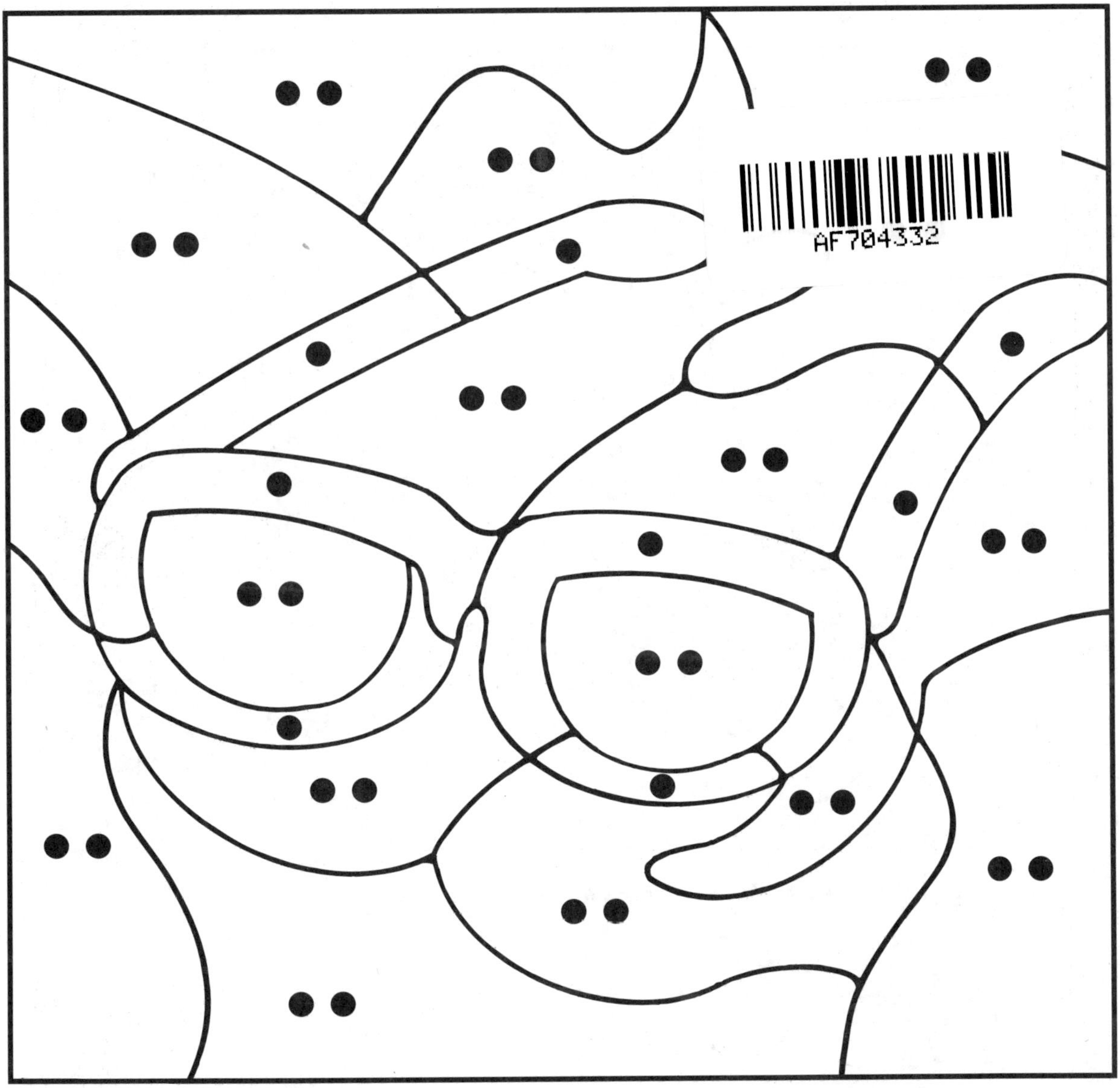

How Many Pets?

Find the pets and color them brown.
Color the rest of the picture.

Circle to show how many.

 1 2

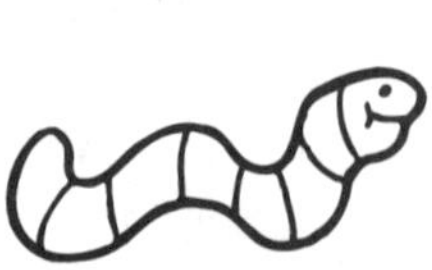 1 2

Rat-a-tat-tat!

Color to find the hidden picture.

▲▲ = orange ▲▲▲ = blue

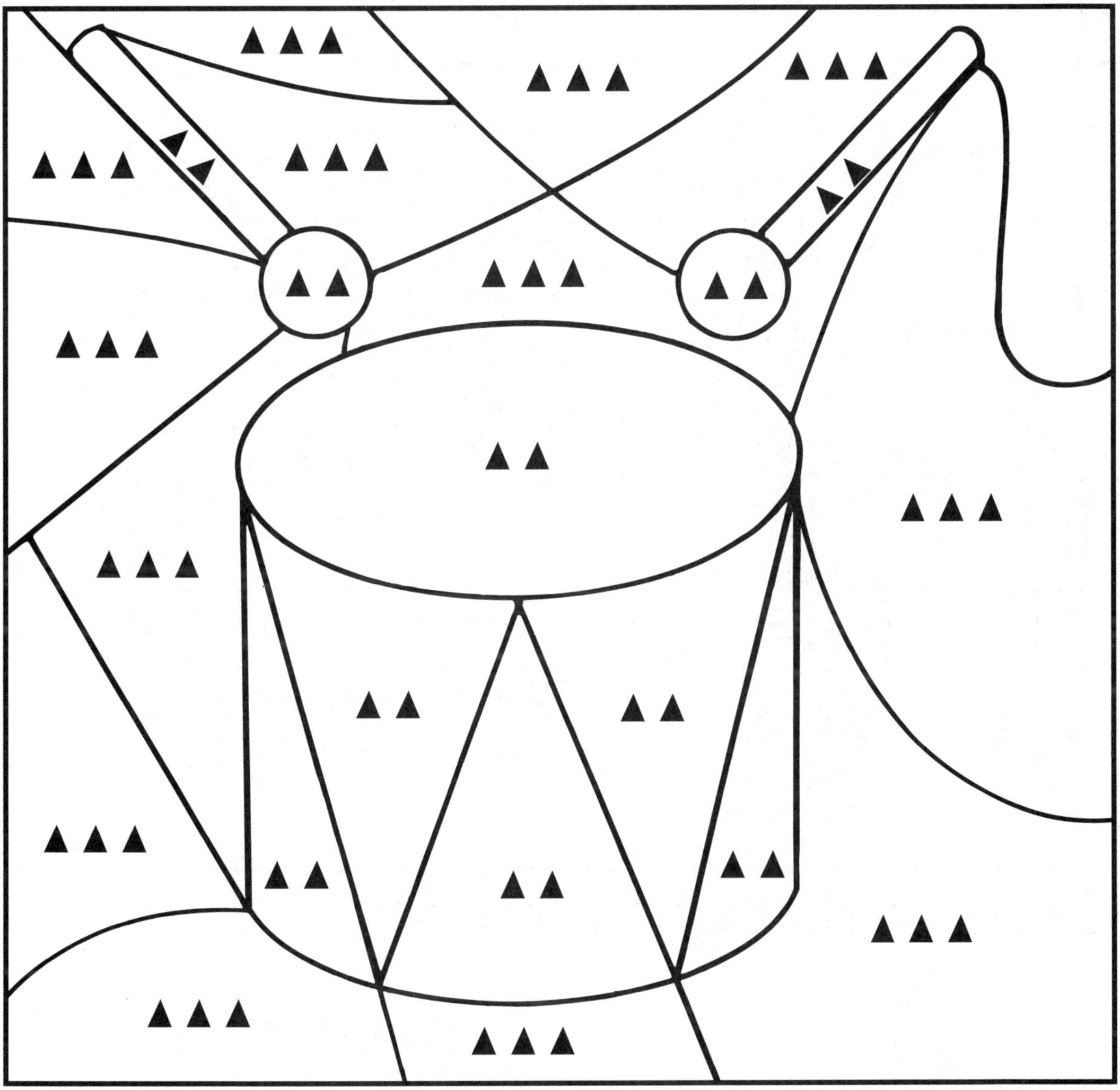

In the Playroom

2 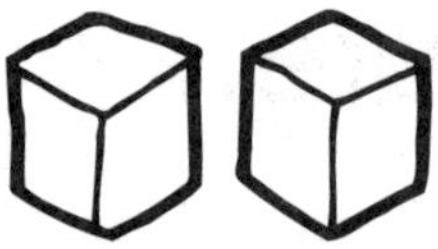 3 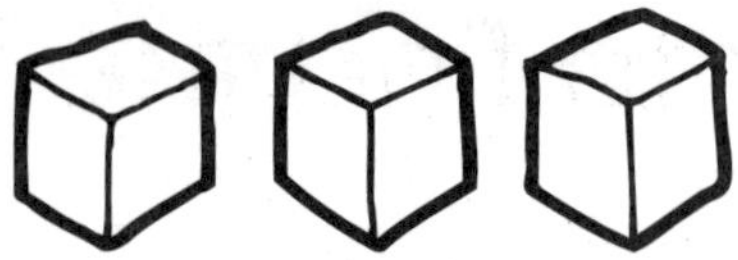

Find the toys and color them red.
Color the rest of the picture.

Circle to show how many.

 2 3  2 3

What Waves in the Wind?

Color to find the hidden picture.

★★★ = purple

★★ ★★ = blue

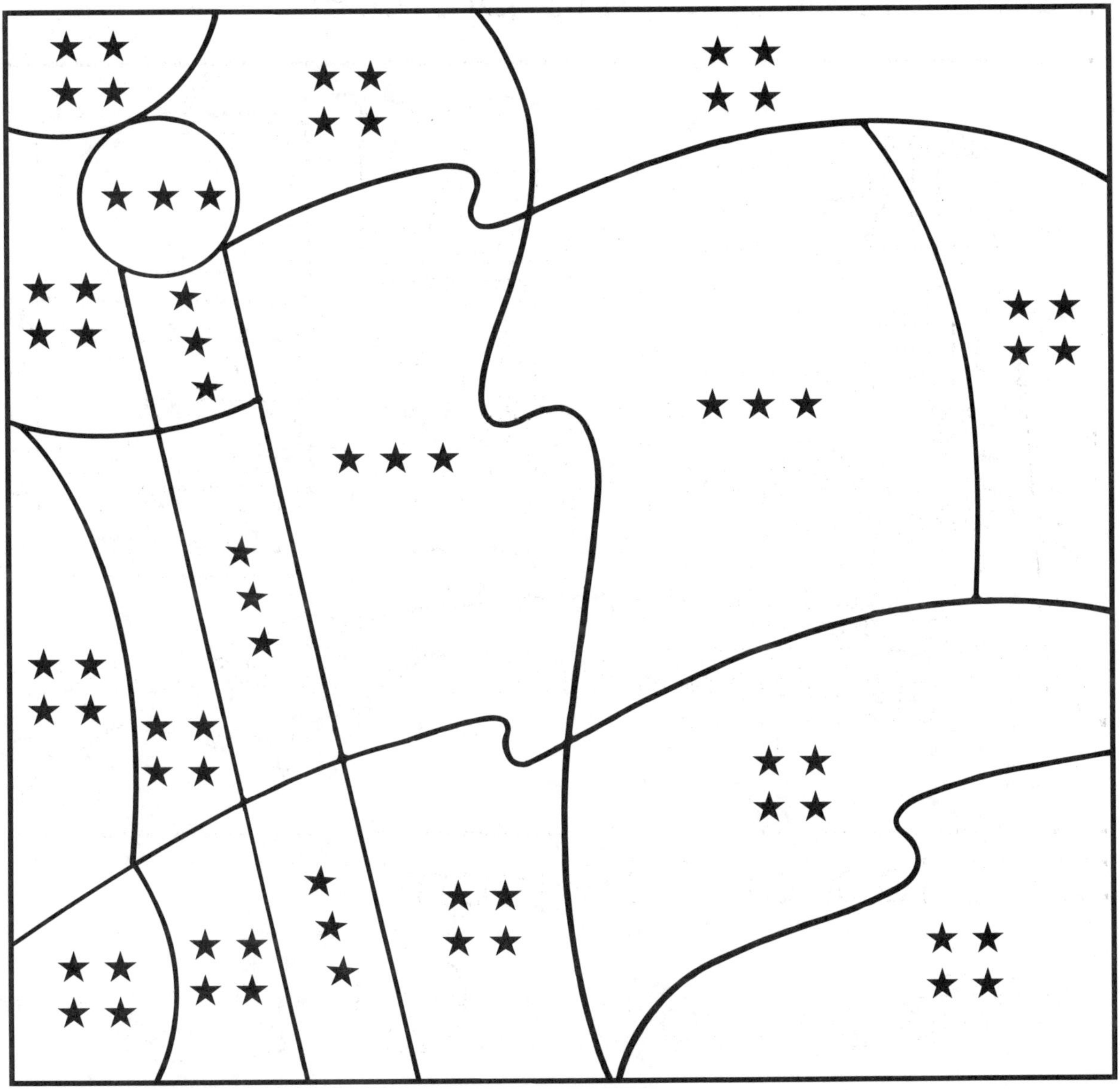

Insect Safari

3  4 

Find the insects and color them red.
Color the rest of the picture.

Circle to show how many.

 3 4  3 4  3 4

On the Farm

Color to find the hidden picture.

 = red  = green

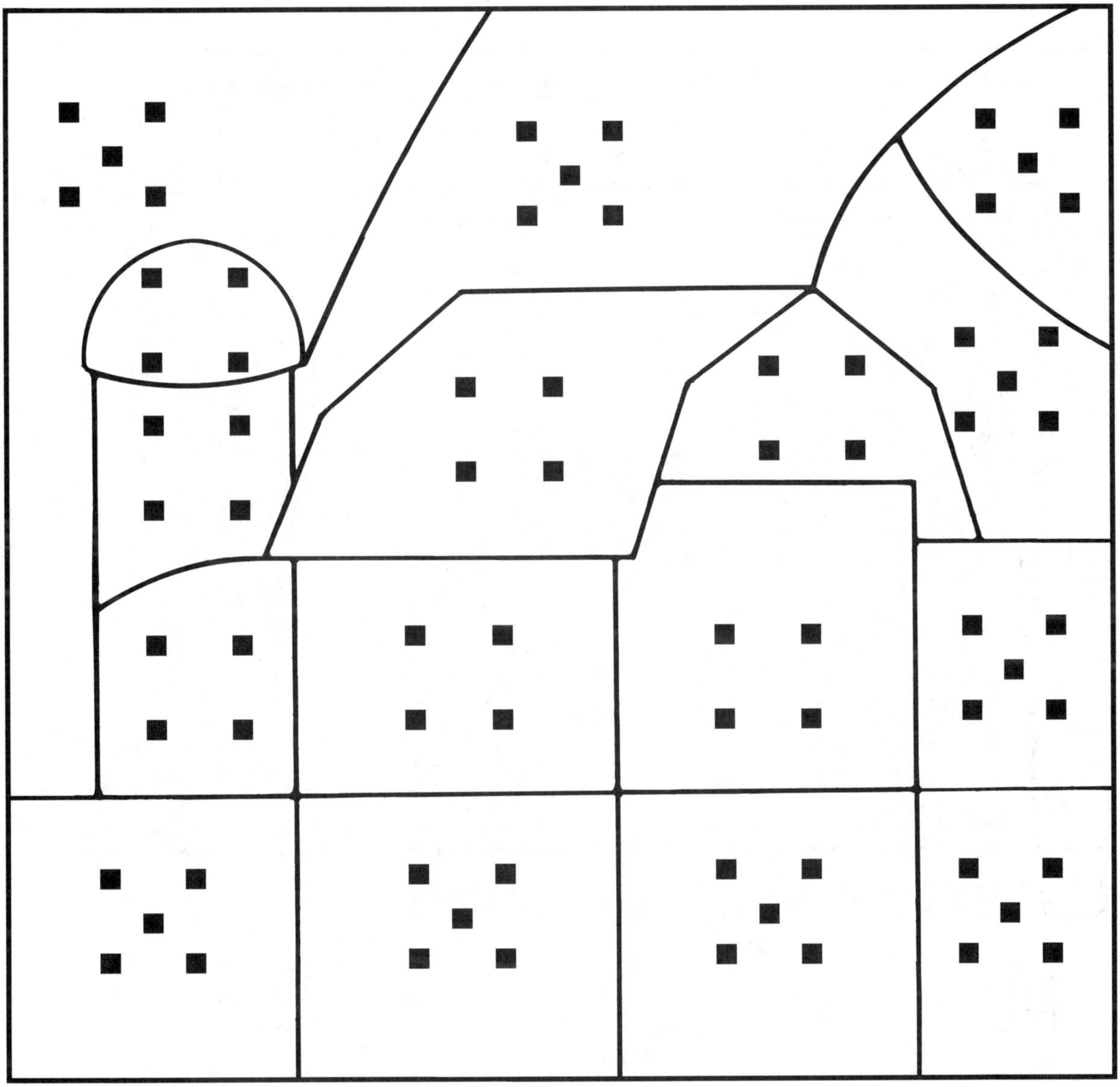

At the Aquarium

4  5 

Find the sea animals and color them yellow. Color the rest of the picture.

Circle to show how many.

 4 5  4 5  4 5

Sailing High

Connect the dots from **1** to **5**. Color.

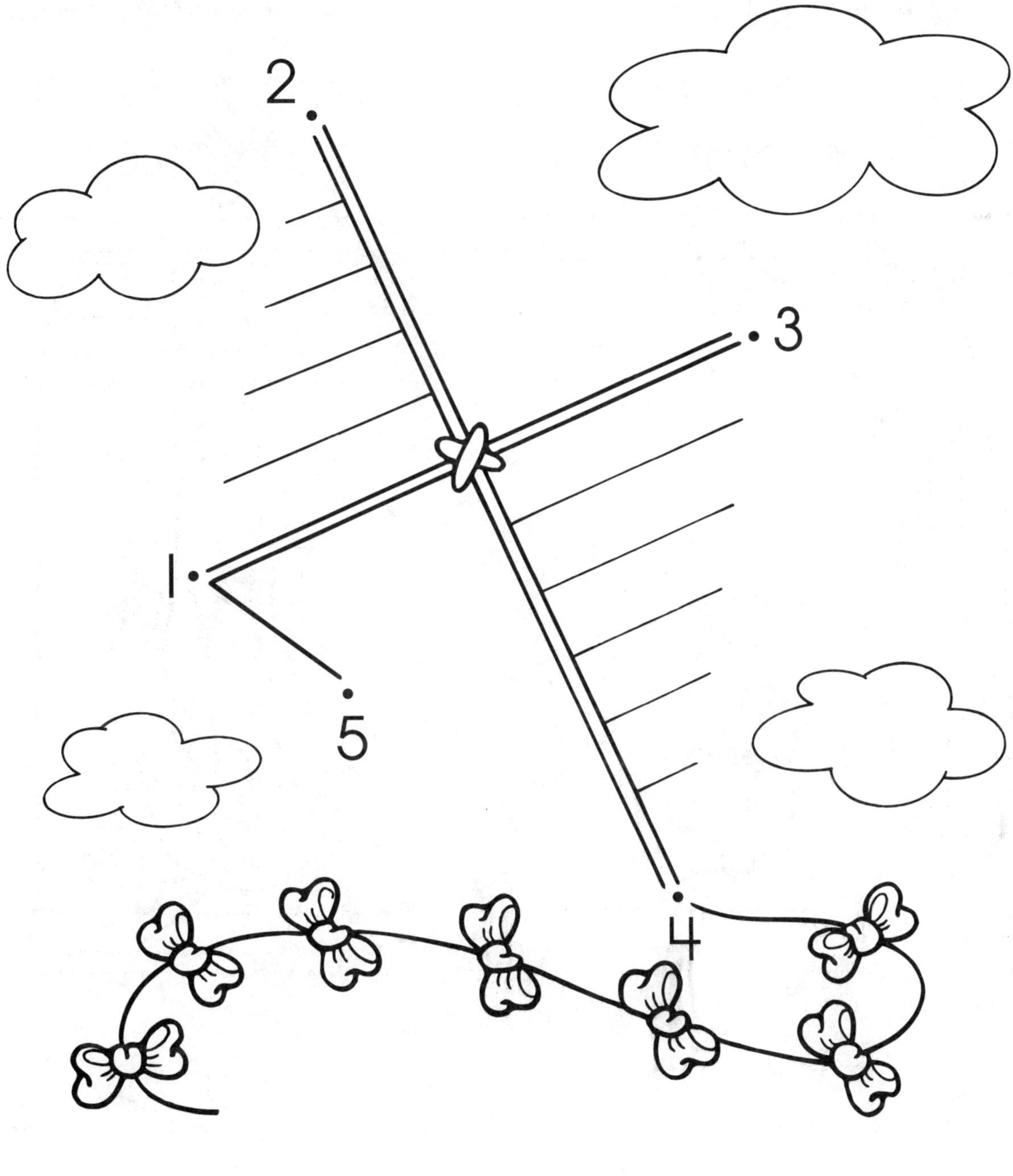

Party Time

Find the numbers **1** to **5** in the hat.
Color them.

Triangle Fun

Count the triangles. Color them.

Circle to show how many triangles you found. 1 2 3 4 5

Something for Your Feet

Color to find the hidden picture.

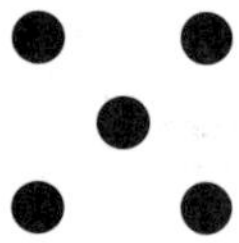 = brown 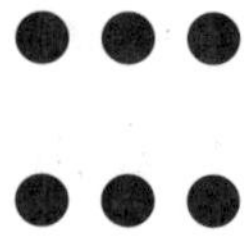 = purple

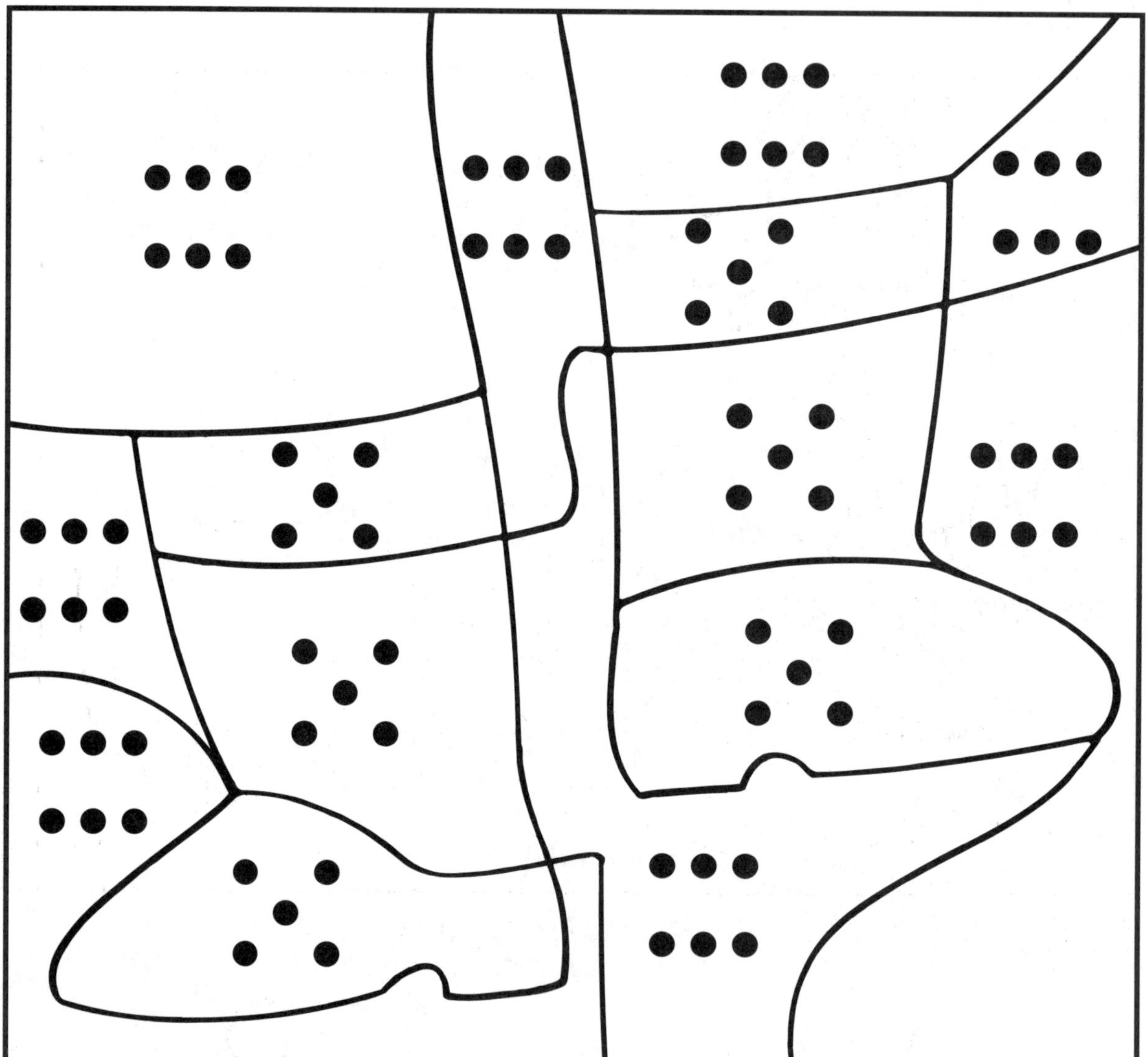

Treetop Friends

5  6 

Find the animals and color them brown.

Circle to show how many.

 5 6  5 6

Fun in the Pool

Color to find the hidden picture.

XXX
XXX = green

XXX
X
XXX = blue

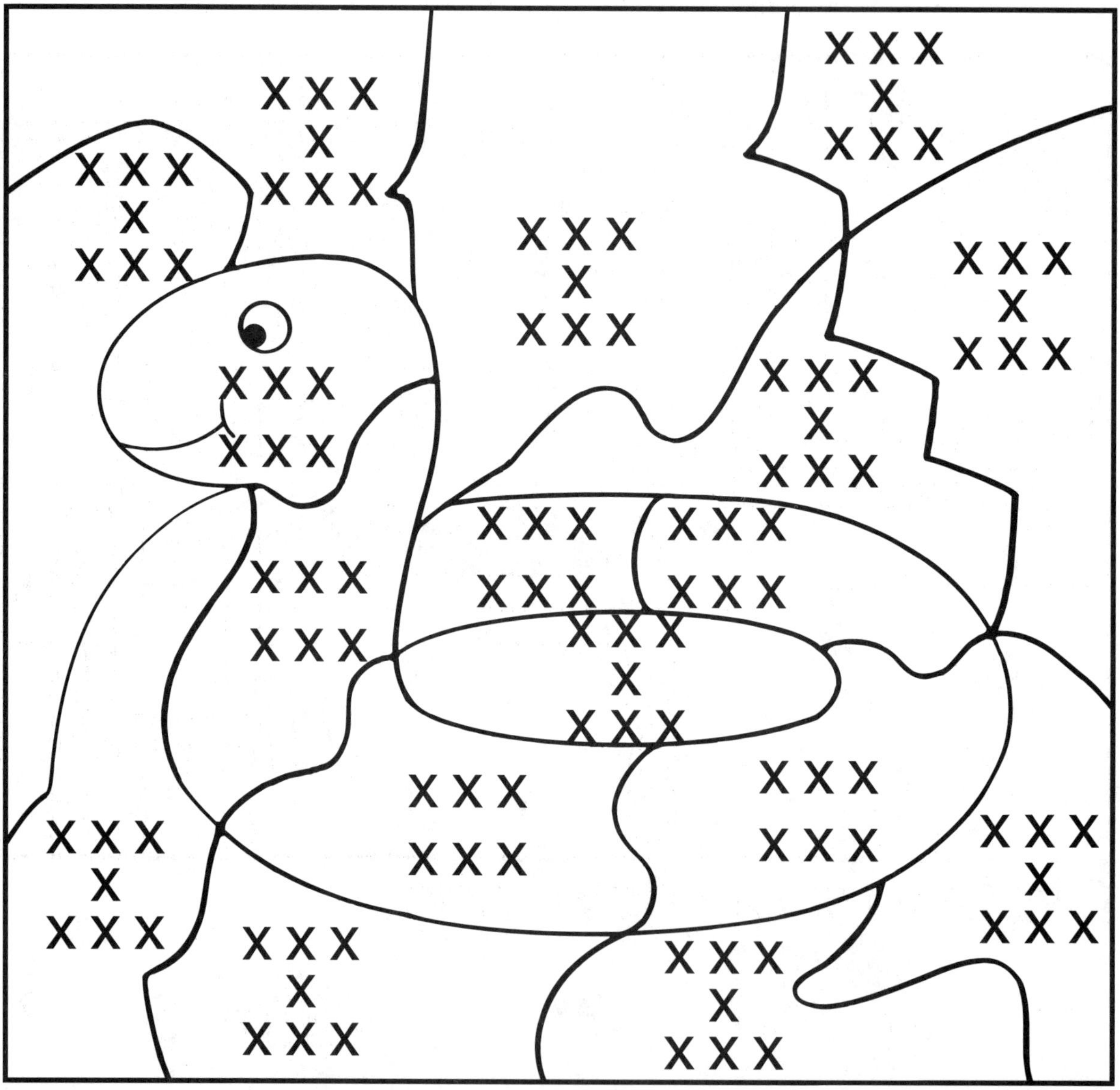

At School

6  7 

Find the books and crayons. Color them red. Color the rest of the picture.

Circle to show how many.

 6 7  6 7

Ding, Dong!

Color to find the hidden picture.

▲▲▲
▲
▲▲▲ = yellow

▲▲▲▲
▲▲▲▲ = red

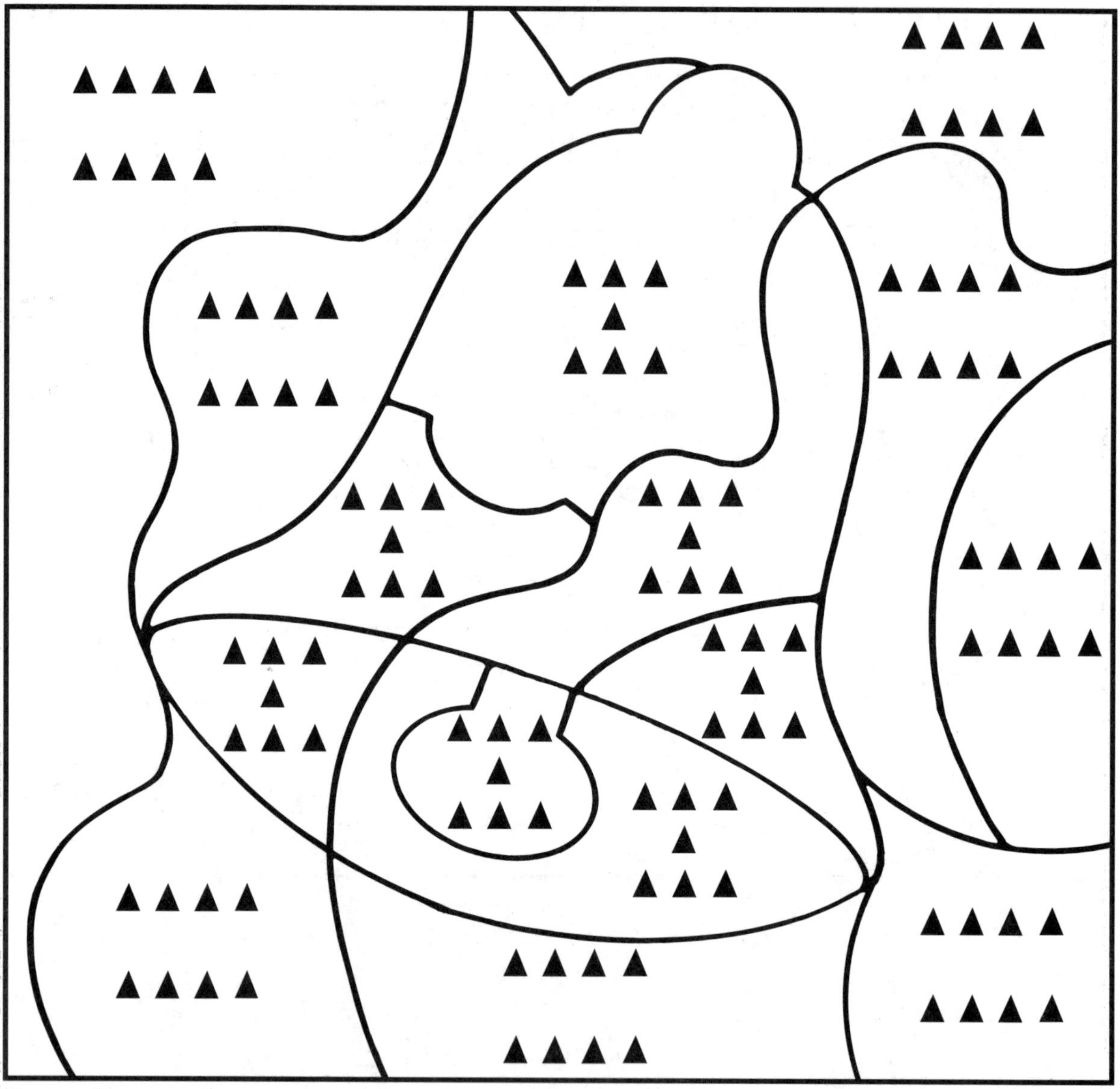

Batter Up!

7 

8 

Find the balls and bats. Color them orange. Color the rest of the picture.

Circle to show how many.

 7 8  7 8

A Long Drop

Connect the dots from **1** to **8**. Color.

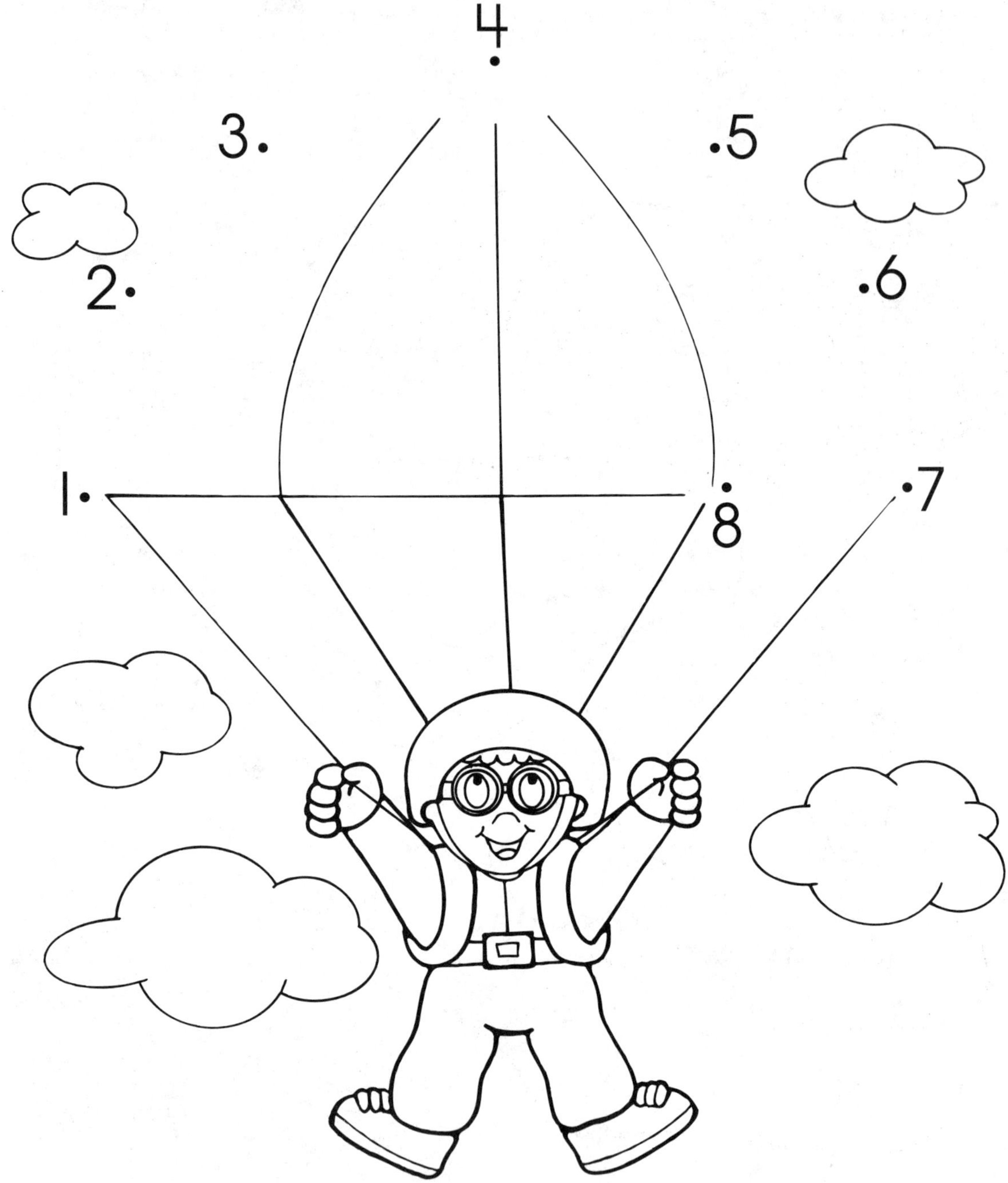

Bo Peep's Sheep

Help Bo Peep find her sheep.
Color the path in order from **1** to **8**.

Make a Pattern

Color the spaces on the umbrella.
The box tells you what colors to use.

yellow	–	**1**	**3**	**5**	**7**
purple	–	**2**	**4**	**6**	**8**

A Bright Light

Color to find the hidden picture.

★★★★
★★★★ = yellow

★ ★ ★
★ ★ ★ = black
★ ★ ★

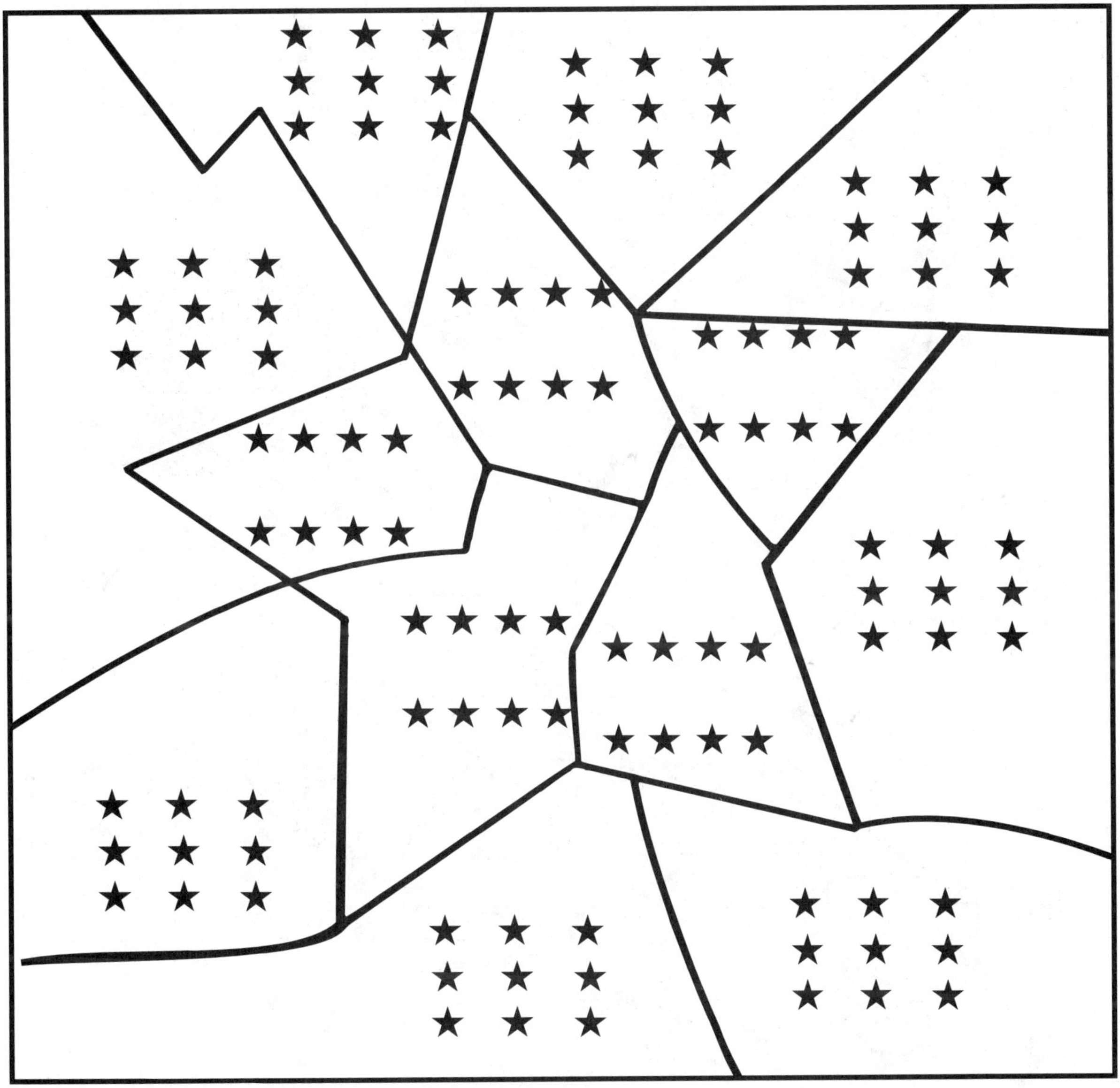

Let's Race!

8  9 

Find the acorns and leaves. Color them brown. Color the rest of the picture.

Circle to show how many.

 8 9  8 9

A Sharp Tool

Color to find the hidden picture.

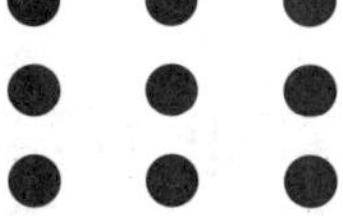 = orange  = green

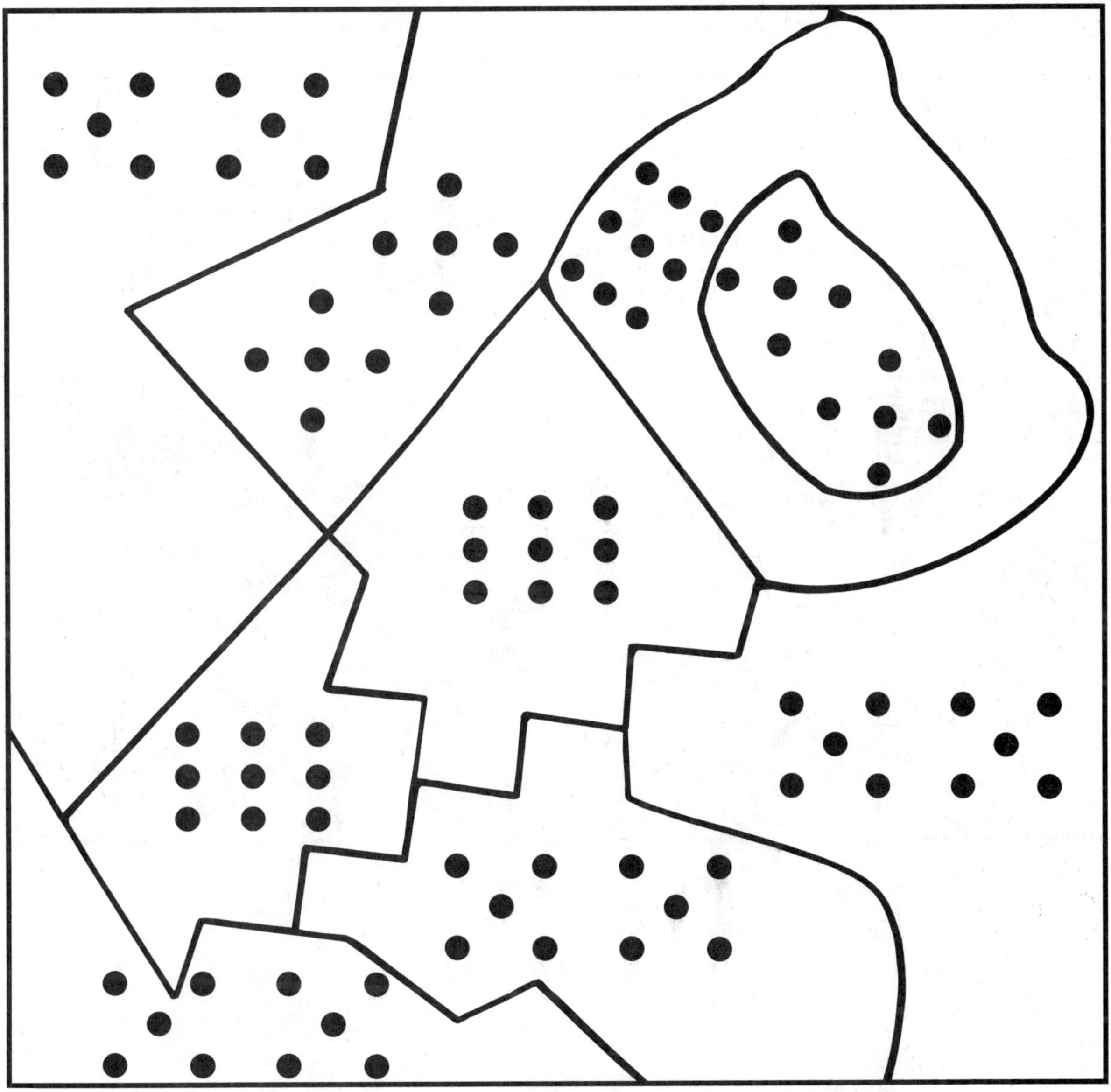

Time for Dinner

9 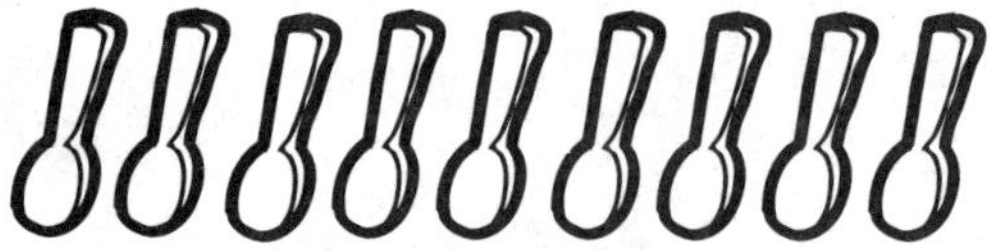 10 

Find the knives and forks. Color them blue. Color the rest of the picture.

Circle to show how many.

 9 10  9 10

Cool and Sweet

Connect the dots from **1** to **10**. Color.

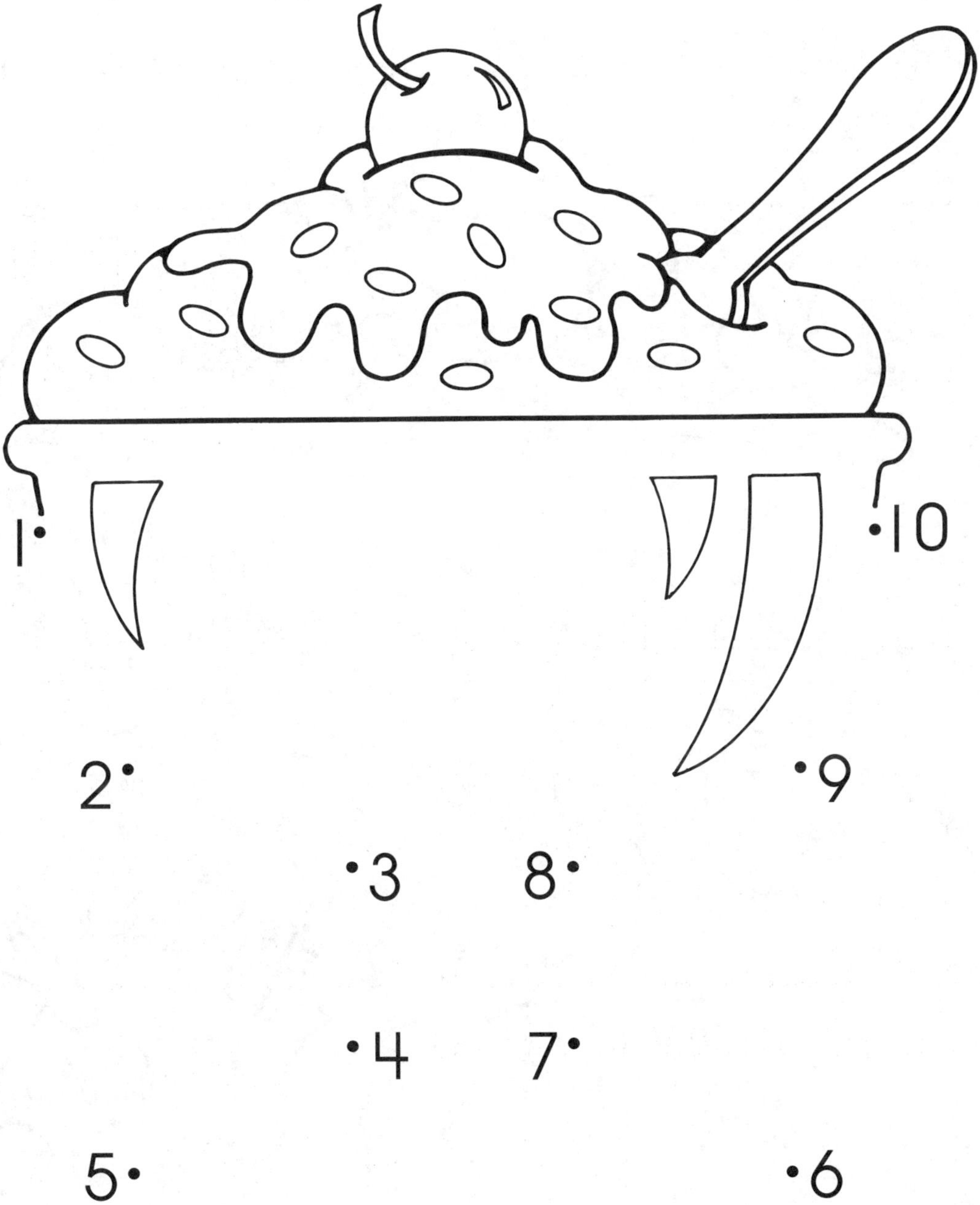

Number Hop

Help the frog get to his friend.
Color the path in order from **1** to **10**.

Cloud Watch

Find the numbers **1** to **10** in the picture. Color them.

1 2 3 4 5 6 7 8 9 10

How Many Squares?

Count the squares. Color them.

Circle to show how many squares you found.

1 2 3 4 5 6 7 8 9 10

A Wiggly Fellow

Color to find the hidden picture.

1 = green **3** = red **6** = blue

A Good Rocker

Color to find the hidden picture.

2 = yellow **4** = brown **7** = green

A Perky Talker

Color to find the hidden picture.

3 = green **5** = orange **8** = red

Spring Blossoms

Color to find the hidden picture.

5 = yellow **7** = green **10** = blue

What Did Ben Make?

Connect the dots from **1** to **10**. Color.

Make a Wish

Help Kelly get to the wishing well.
Color the path in order from **1** to **10**.

Rabbit's Garden

11 

Find the carrots and color them orange. Can you find **11** carrots in all?

Collecting Shells

12 

Find the shells and color them purple.
Can you find **12** shells in all?

A Big Scooper

Color to find the hidden picture.

10 = orange **11** = yellow **12** = blue

Peekaboo!

Connect the dots from **1** to **12**. Color.

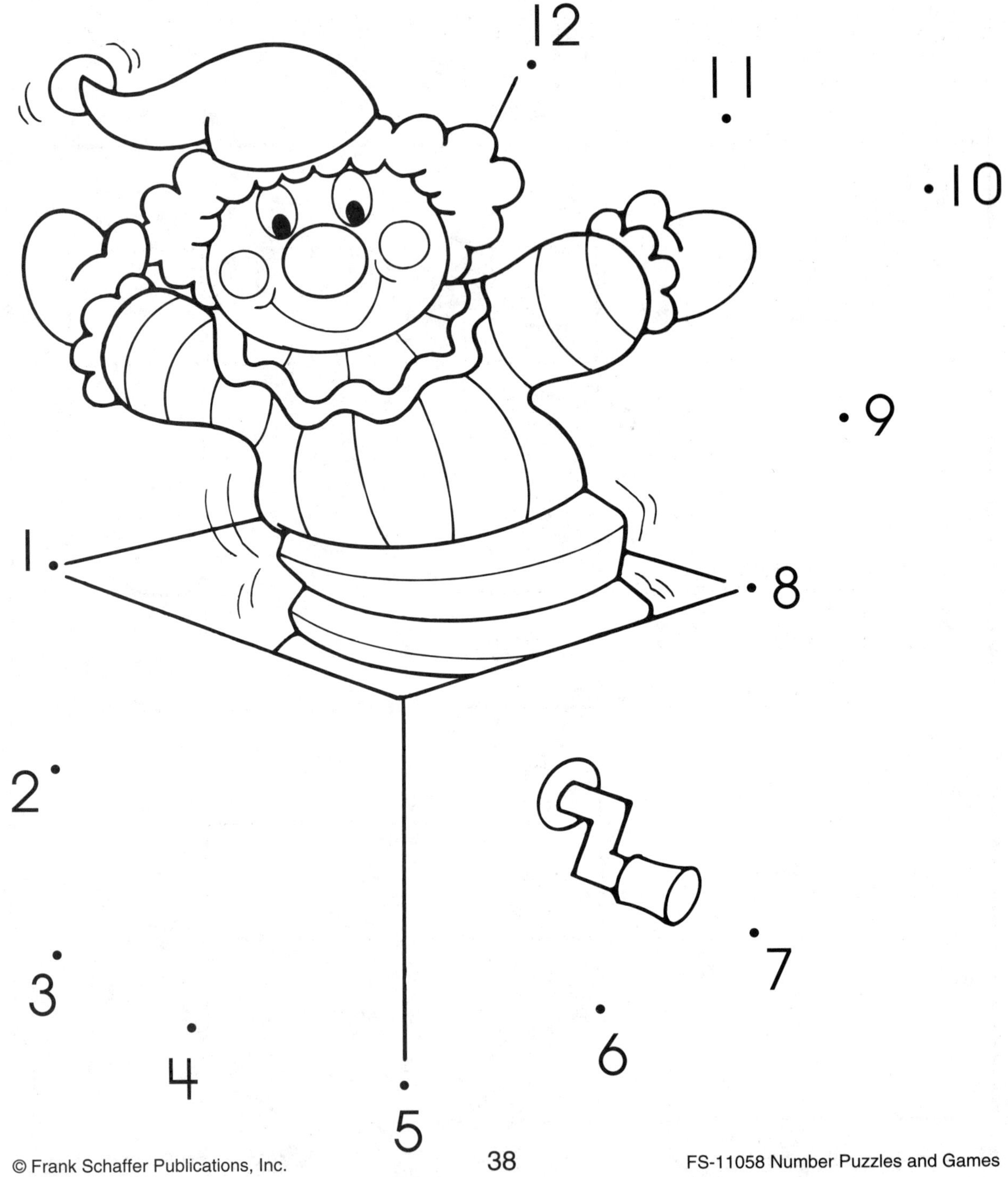

Scruffy's Bone

Help Scruffy get to the bone.
Color the path in order from **1** to **12**.

	1	2	3	5	2
			4		1
			5		1
5	4	7	6	1	1
6		8			
2	3	9	10	11	
				12	

All Dressed Up

13 

Find the rings and color them red.
Can you find **13** rings in all?

What Likes Honey?

Color to find the hidden picture.

11 = yellow **12** = brown **13** = green

Washing Day

14 

Find the socks and color them blue.
Can you find **14** socks in all?

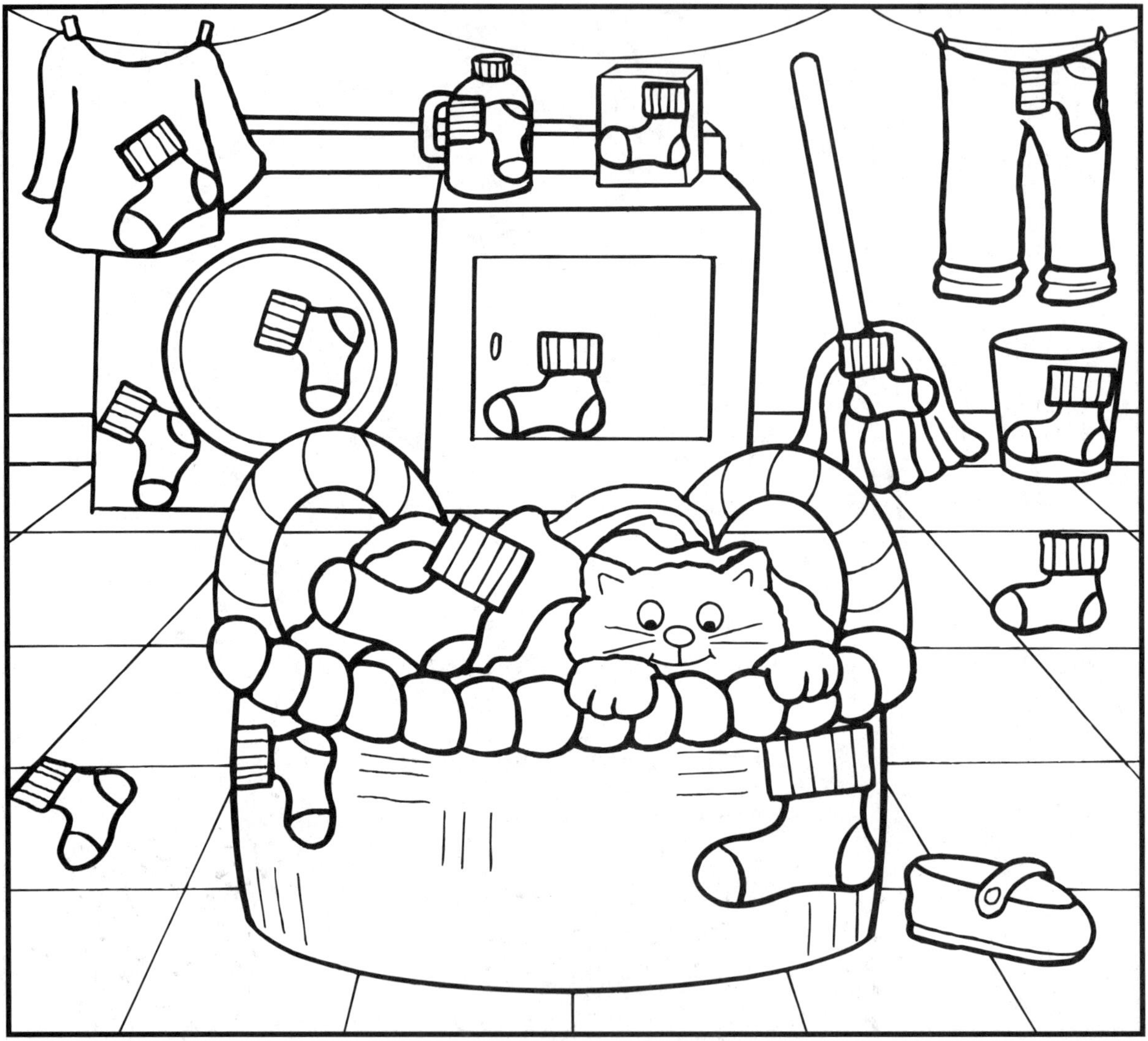

A Nutty Snack

15 

Find the peanuts and color them brown. Can you find **15** peanuts in all?

A Fast Spinner

Color to find the hidden picture.

13 = red **14** = yellow **15** = purple

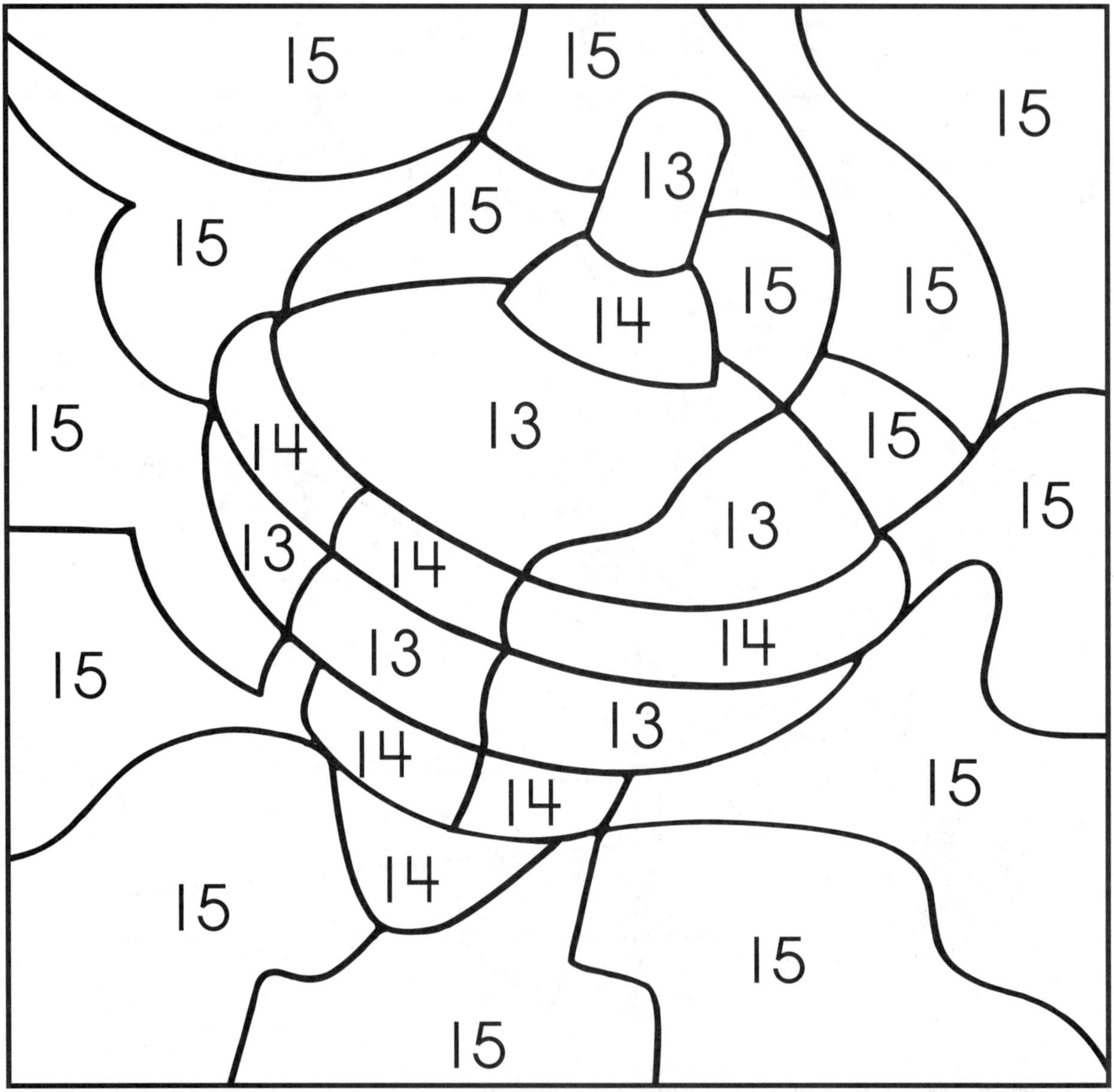

What Can Make Wind?

Connect the dots from **1** to **15**. Color.

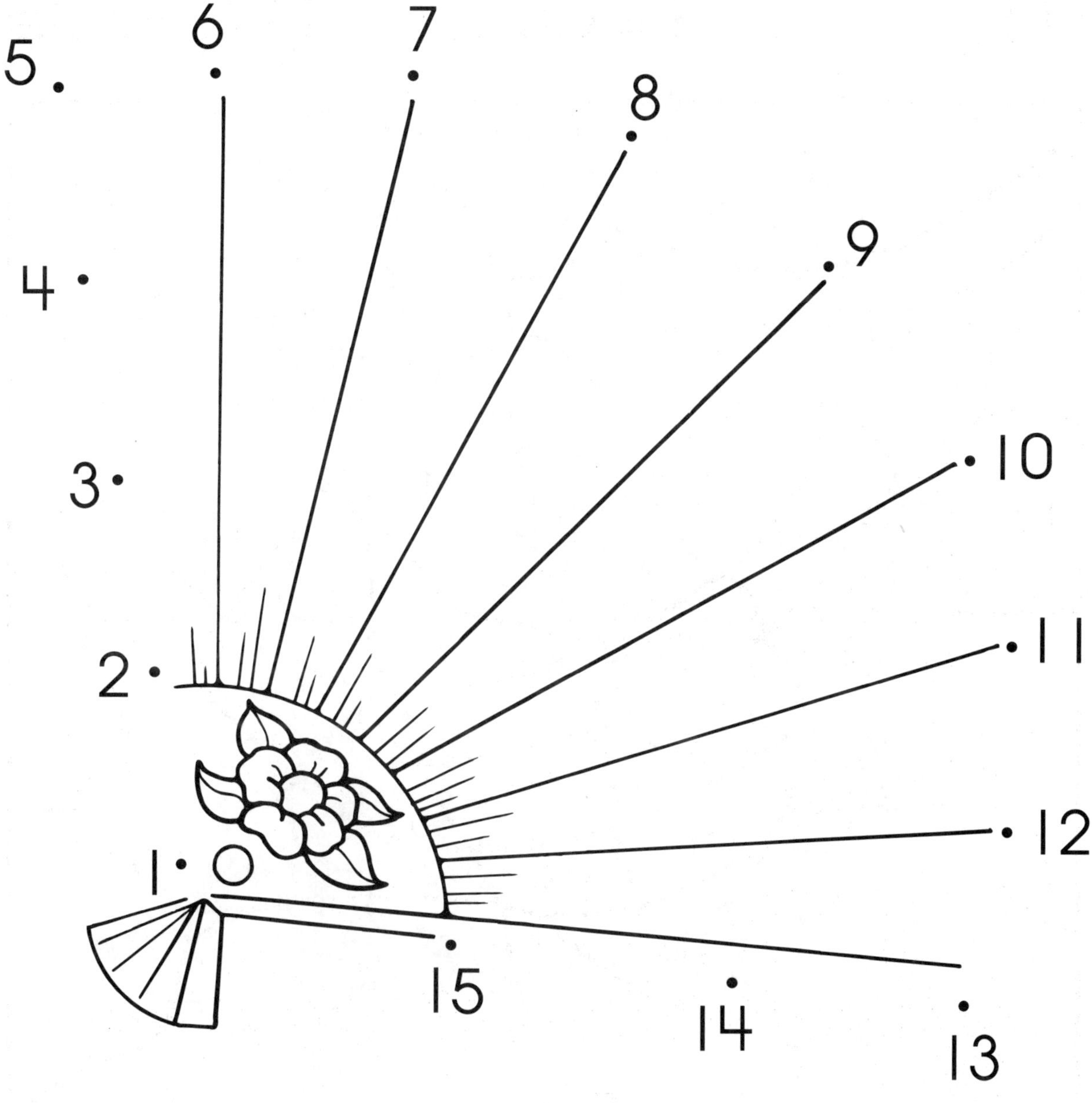

Let's Go Rowing

Help the children get to the rowboat.
Color the path in order from **1** to **15**.

Gathering Eggs

16 

Find the eggs and color them yellow.
Can you find **16** eggs in all?

Time to Brush

17 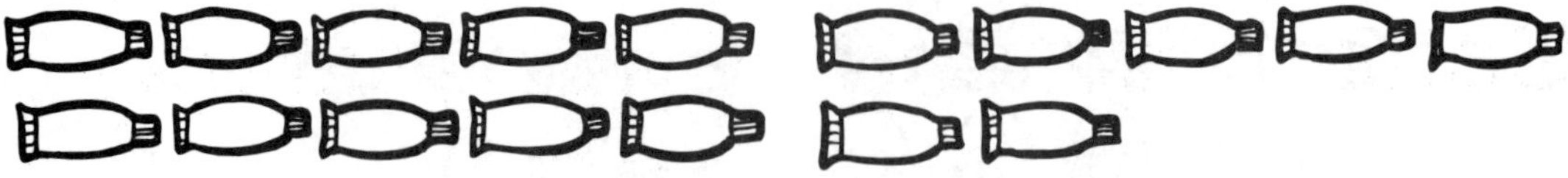

Find the toothbrushes and color them blue. Can you find **17** toothbrushes in all?

A Hungry Nibbler

Color to find the hidden picture.

15 = orange **16** = brown **17** = green

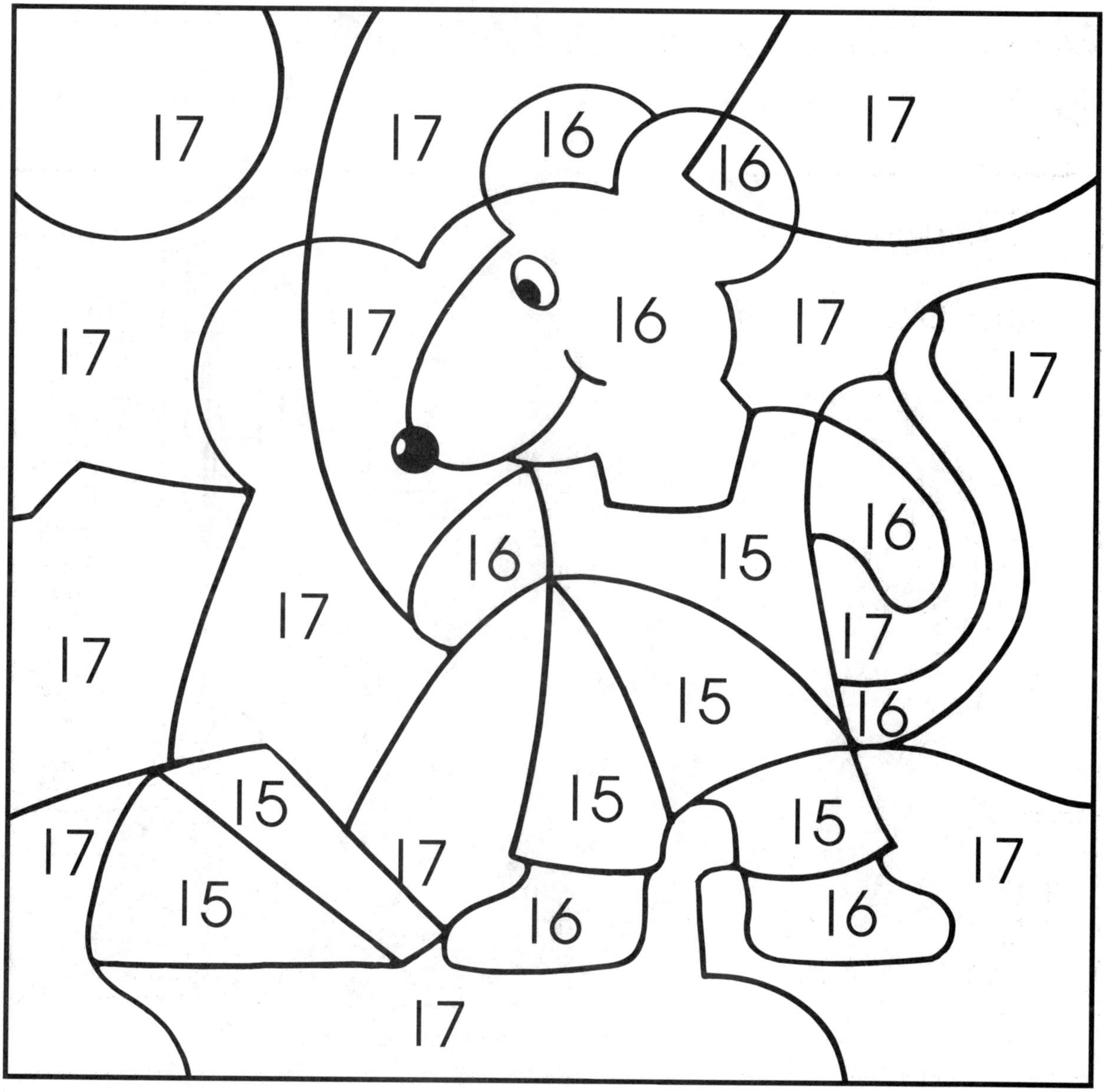

Going Out

18 

Find the keys and color them blue.
Can you find **18** keys in all?

Let's Paint

19 

Find the paintbrushes and color them green. Can you find **19** paintbrushes in all?

Birthday Fun

20 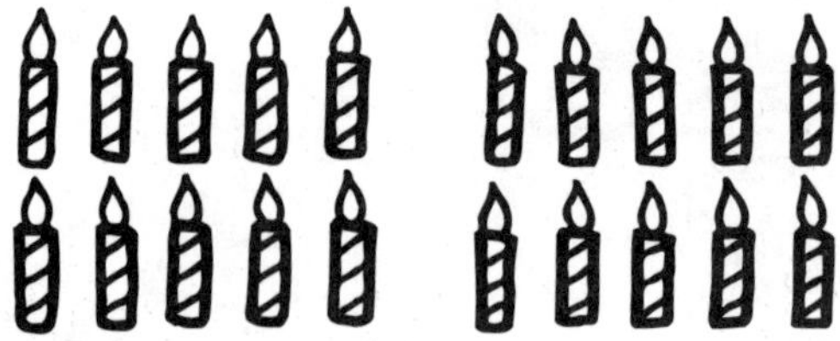

Find the birthday cupcakes. Color them red. Can you find **20** cupcakes in all?

A Splash of Color

Color to find the hidden picture.

18 = red **19** = blue **20** = yellow

Who Wants a Cool Drink?

Connect the dots from **1** to **20**. Color.

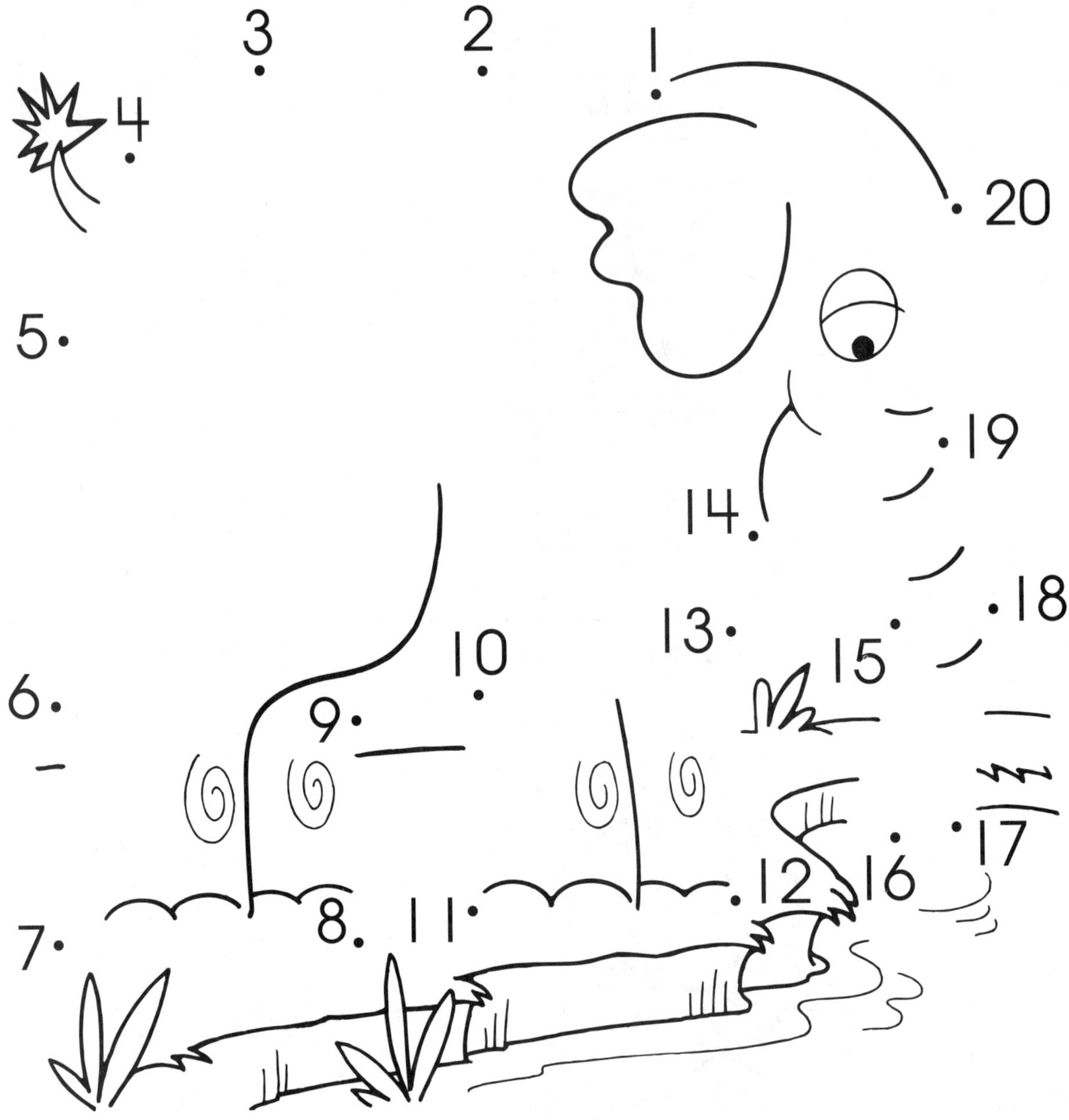

What's in the Cupboard?

To find the mystery number, color the spaces with these numbers purple.

9 5 8 10 18 14 3 16 20 12

Circle the mystery number. 4 9 17

School Days

Help the children get to school.
Color the path in order from **1** to **20**.

		1	2	3	4	5
			5			6
10	9	8	15	14	13	7
11						8
12	7	13	12	11	10	9
13		14		6		
14		15		19	20	
12	11	16	17	18